Ginkgo Biloba

Stephan Brown

For my son, Corey, my great joy, from his "nuts 'n' twigs" dad.

CONTENTS

Introduction 2
Why Is Ginkgo So Popular? 2
A Love for Ginkgo Is Born...4
The History of Ginkgo 6
Growing Your Own Ginkgo 7
Propagating New Ginkgos..10
Growing Ginkgo as a Cash Crop...13
Commercial Standardized Extracts 16
Making & Using Your Own Ginkgo Tincture 17
My Favorite Ginkgo Recipes 22
Conclusion 30
Resources 31

Introduction

The ancient and venerable ginkgo tree *(Ginkgo biloba)* is the oldest tree species on planet Earth. Its fossil records go back a staggering 200 million years! Ginkgo stood tall as dinosaurs brushed past, and pterodactyls may have lunched on lizards sheltering in its leafy arms. With its unique, fan-shaped leaves; silvery gray bark; and strange, sometimes random branching pattern, ginkgo is particularly memorable among the world's great trees.

But ginkgo is more than a tree. Because of its great age in the unimaginable spectrum of geologic time, ginkgo epitomizes persistence — fighting the odds and surviving. It offers a model of stamina and endurance for us to aspire to through the difficult challenges that are encountered in this life. As a medicinal herb, ginkgo is best known for its improvement of short-term memory, but it is also effective in treating allergies, depression, poor circulation, and many more conditions. What a magical tree!

The Native American people sought wisdom from — and sometimes would wrap their arms around — the largest tree they could find, believing that those of great age also had great knowledge and spiritual strength. I have said countless times, "You can hug a tree — or you can hug a ginkgo!" With its amazing history, would not ginkgo, of all the trees on earth, be the most knowledgeable and most powerful? I believe that ginkgo is both the teacher and the protector of all the other plants on earth.

Why Is Ginkgo So Popular?

Ginkgo is revered by gardeners as an ornamental tree, and it is also one of the most popular and widely researched medicinal herbs in the world. Extracts of the leaves seem to have a somewhat unique ability to facilitate the body's peripheral circulation and the microcirculation to the brain, including the ocular area. It is thus quite useful in helping those with cold hands and feet, certain neuropathies, poor short-term memory, poor removal of natural metabolic wastes, and any condition associated with cerebral ischemia (a lack of oxygen and glucose in the brain). Increased cerebral circulation enables the blood to bring more oxygen and glucose to the brain and to carry away the metabolic wastes that would poison the brain if not promptly removed.

In other ways (the technicalities of which are more adequately described in other sources), ginkgo has a tonifying, or invigorating, effect upon the central nervous system. It is useful in conditions of age-related cognitive decline (dementia), poor memory, and decreased learning abilities. Ginkgo has also proved beneficial in cases of head injury, early Alzheimer's disease, stroke, impaired hearing, tinnitus, eye problems, depression, allergies, intermittent claudication (achy legs), and cellular damage from free-radical oxygen molecules.

With all its many applications, ginkgo causes only rare instances of adverse reactions or side effects. Usually an upset stomach is the only complaint. It's no wonder that ginkgo is such a popular herb!

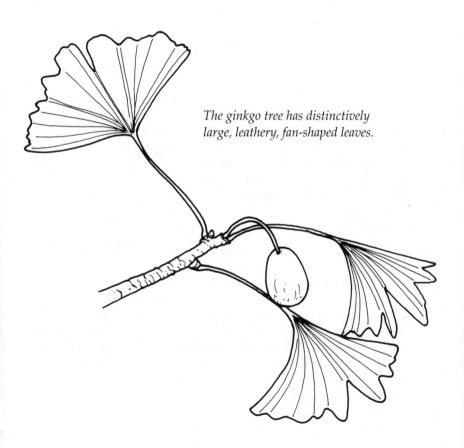

The ginkgo tree has distinctively large, leathery, fan-shaped leaves.

A Love for Ginkgo Is Born

I am a self-admitted ginkgo fanatic. My love affair with ginkgo trees began back in 1973. After five years of mowing lawns and performing basic landscaping services, I used $3,000 of my own money, some money I borrowed from my family, and a second mortgage to purchase land that soon became my own plant nursery in Brewster, Massachusetts, on the Cape Cod peninsula.

I became a regular customer of the large wholesale nurseries in Newport, Rhode Island. I would drive my old pickup truck and homemade trailer the two hours from Brewster to Newport to pick up stock. Most of the nurserymen in that area were Dutch. They had immigrated in the early part of the century and began to ply the trade that they knew best: growing ornamental plants. Being young, energetic, and insatiably curious, I barraged the old Dutchmen — the Hoogendoorns, Kempenaars, Vander Pols, and Vaniceks — with questions about raising, feeding, pruning, and propagating these newfound plant friends of mine.

One day I was looking through the rows of plants at Hoogendoorn's nursery — the junipers, rhododendrons, lilacs, and others — when suddenly, there in front of me were three or four trees the likes of which I had never before seen. They captured my complete attention. They were strange, beautiful, and totally unique with their fanlike leaves, a silvery bark that was ridged yet smooth, almost horizontal branches, and leathery-textured leaves. Looking more closely at a leaf, I saw that the veining started at the petiole and radiated straight out to the edge of the leaf.

"George!" I gasped in amazement to my horticulturalist host. "What *is* this? I've never seen anything like it before!"

"That's a ginkgo," he offered back. "It's the oldest tree on earth. Dinosaurs were around when this tree was growing."

Thus began my obsession with the fascinating ginkgo. For many years, I seemed to enjoy this passion alone, as few other people were interested in the ginkgo as an ornamental tree. More recently, as the interest in medicinal plants has skyrocketed, more and more people have read the influx of scientific data confirming ginkgo's health benefits and are now seeking it out. Today I sell thousands of these beautiful trees to people all over the country, and I always include a plethora of serious prose and puckish poetry with each shipment.

Confessions of a Ginkgo Fanatic

Driven by my ginkgo passion, in 1999 I journeyed to the opposite side of the earth, to China, where the oldest ginkgo trees on earth grow. My pilgrimage brought me to three ancient ginkgo trees an hour's drive from Nanjing, China.

Silhouetted against a somber, gray China sky, they stood like giants, dwarfing all surrounding life-forms as their ancient ancestors had done for 200 million years before them. These were the masters, the omniscient, the ancient ones, and they loomed above a small Buddhist temple at the end of a dirt path. These were three of the very few remaining indigenous ginkgos left on earth.

The trees had been adorned with red ribbons (tradionally a symbol of good luck) and personal mementos, and I could see from the ashes that incense had been burned many times at their base, the fragrant smoke lifting through their huge, protective branches into the air of a thousand civilizations.

I stood transfixed for many minutes. It was overpowering to approach these wise ones after so many years of waiting for and dreaming of this meeting. It was humbling to be able to touch these oldest ancestors of the ancient clan of trees. There I was, only one of billions of humans and other living creatures who had gone before me, a mere blink in the spectrum of time, standing before three huge trees whose ancestors had communed with dinosaurs in the pure, oxygen-rich atmosphere of a prehistoric planet. There I was, in the presence of the oldest of old. There are some feelings that words can never describe. I was in that indescribable space of complete silence and stillness where the spirit feels at one with the momentum of the universe.

The History of Ginkgo

In prehistoric times, ginkgo grew over most of the planet. In the United States, for example, we can find petrified ginkgo logs at the Ginkgo Petrified Forest State Park in Vantage, Washington. Sadly, populations across the globe died out; in the modern geologic age, which began a few thousand years ago, indigenous plants can be found only in one area of China.

One theory about why ginkgo has survived for so long suggests that Buddhist monks cultivated and protected the ginkgo at their monasteries. Although some very old trees do stand near monasteries or temples, this explanation may be a bit romanticized. What is very explicable, though, is why the ginkgo tree now graces most temperate countries of the world: Dutch, English, Portuguese, and French explorers and traders ventured into the Orient in the 1700s and 1800s. We know that they returned with spices, silk, and precious stones, but they also introduced many new plant species to Western countries. Among these was ginkgo, and the first recorded planting was at Kew Gardens in England in 1732.

The oldest ginkgo tree planted in the United States dates to 1784 and can still be seen today outside of Philadelphia. There are many very old trees throughout the northeastern United States; some remarkable trees can be found at Union College at Schenectady, New York, and at the University of New Hampshire at Durham. New York City and Montpelier, Vermont, are also home to old ginkgos. The oldest that I have found on Cape Cod is on Main Street in Hyannis, between the Cape Cod Bank and Trust and the *Cape Cod Times* buildings.

Seen Any Old Ginkgos?

I am very curious about how, where, and when ginkgo spread out from its indigenous populations in China and Japan since the 1700s. If you know of any very old ginkgo trees in North America or worldwide, I'd love to hear from you. Any history about these old trees would be welcome as well. I can be reached at my company, Great Cape Herbs (see the resources on page 31 for contact information).

Growing Your Own Ginkgo

Ginkgos are very hardy, easy to grow, and will thrive in a huge range of temperature, soil, light, and water conditions. They are particularly useful as city trees because they are unusually tolerant of city-generated pollution in the air and water. They prefer a sandy loam, although they will grow in just about any soil. They seem to be able to survive with only spare amounts of water, as indicated by those city ginkgo that grow in the tiniest of planting holes in stone, concrete, or asphalt sidewalks. They can easily spread to 50 feet in width and attain a height of 100 feet.

Growing at a Glance

CLIMATE PREFERENCE:
Zones 4–9

SOIL REQUIREMENTS:
Prefers sandy loam but will tolerate most soil conditions.

LIGHT REQUIREMENTS:
Partial shade

WATER REQUIREMENTS:
Spare

SPACING REQUIREMENTS:
Space ornamental specimen ginkgos 20 feet from other full-grown trees. Space gingko crop trees 6 to 8 feet apart, and keep pruned to under 8 feet for ease of harvesting.

Suitable Climates

Ginkgo's naturally preferred climate is Zone 6 or 7. However, I have many customers who are growing ginkgo successfully in Zone 4, which experiences temperatures as low as -25°F.

I knew that gingko grows as far south as northern Florida, but one day a participant in one of the weekly "weed walks" at my farm told us that she had seen a ginkgo at the Fairchild Botanic Garden in Miami. So now I tell people that this superstar species can grow from Miami to Montreal!

The majority of the old-growth trees in China grow in Jiangsu Province, which lies between Beijing and Shanghai. A similar climate in the United States might be the area from northern South Carolina to New Jersey. Ginkgo will also grow up and down the West Coast as well as at fairly high elevations; an herbalist friend in Colorado has them growing at 6,000 feet above sea level.

Ginkgo may have a difficult time growing in arid regions. I have heard little or nothing of ginkgo growing in the American Southwest, for example.

At Home in the City

In the town of Hyannis here on Cape Cod, the largest ginkgo tree that I have found is growing out of an asphalt driveway between a bank and a newspaper office on Main Street. Its roots are completely covered for hundreds of feet in all directions by asphalt. You will probably see this ginkgo tenacity repeated in many cities throughout the world.

Better Medicine Comes from Greater Stress

Optimal growing conditions are most often thought of as those conditions conducive to growing the biggest, lushest tree possible. But if you are growing ginkgo for medicinal quality and not necessarily for leaf quantity, or what's called "biomass," then it is not desirable to pamper your tree.

Horticulture professor Dr. He Shan-an, the recently retired director of the Nanjing Botanical Garden in China, has measured the medicinal quality of ginkgo leaves from trees grown under varying degrees of environmental stress. A synopsis of his findings shows that plants grown at the outer limits of their naturally preferred environmental conditions produce natural chemicals — essential oils — that help them to survive in comparatively harsh conditions. These oils are the medicine that the tree has to offer us.

To analogize, we might look at culinary herbs grown in the harsh sun and rocky soil of the Mediterranean region. These plants produce an abundance of essential oils and waxes that help them survive blazing sun, little rain, and thin soil. If the herbs are grown in rich garden soil and fed and watered frequently, they will grow into fat, lush plants but will not produce the same delicious and fragrant oils as their wild Mediterranean cousins. The same holds true for ginkgo: The greater the stress, the better the medicine.

The catch-22 here is that most farmers want to produce as much "biomass" as possible to realize more profit from more leaves sold at market. But the best leaves come from trees that produce the fewest leaves! So, dear reader, if you want a big, beautiful ginkgo for your front lawn, pamper your tree. But if you want good medicine from your tree, you must be a cruel master! The trick is to create enough stress — via controlling water, sun, soil, and food — to encourage the tree to make good medicine, but not so much that survival is jeopardized. This is the challenge of the grower of good ginkgo medicine.

Ginkgo's Autumn Quirk

One morning in late October many years ago, I awoke and looked out of the big window in the little barn I was living in on my farm in Brewster, Massachusetts. The sun was clear and bright and the October air cool. At the edge of my backyard garden stood Grampa Ginkgo (my oldest tree) in his full autumn glory of bright yellow leaves.

That evening as I returned to the barn, I looked over at Grampa, and my jaw dropped. He was stark naked! I was both confused and amazed, as there had been little wind that day. It was then that I learned that ginkgo is one of only a few plants on the whole earth that drops almost all of its leaves on *one day!* On top of all the other magical things about ginkgo trees, they have a built-in clock somehow connected to the biorhythms of the universe, that tells them which day is the right day to drop their leaves. Is it weather related? A phototropic reaction? Astrologically instigated? We simply don't know.

If you have a ginkgo or live near one in a northern climate (I have been told that this quirk is not as pronounced in warmer latitudes), keep an eye on it this fall. Record when it drops its leaves, how long it takes to do so, what the weather conditions are like, and any other information that you think might be relevant.

Propagating New Ginkgos

Ginkgos are dioecious, meaning that they are single-sex trees; some are male, and some are female. Male and female trees must be within pollination proximity to each other for fertile seeds to be produced.

Ginkgo is more difficult to propagate from cuttings than it is to grow from seed. However, if you want to be guaranteed one gender or the other, use the cutting method. Many people who grow ginkgo as an ornamental tree wish to have a male tree, since the rotting fruits of the female tree have a very disagreeable odor. If you are growing your trees with the intent of using the leaves for medicine, the gender is less important, as you will probably be keeping the trees pruned to under 8 feet tall to allow for easier harvesting of both the leaves and the fruit.

If any of your ginkgo trees are female, you can harvest the fruits and sow them, eat them, or sell them to those who know of their benefits. It is unfortunate that so few Western people are aware that ginkgo nuts can be eaten and can benefit the organs of elimination. Perhaps this bulletin will create a demand for ginkgo nuts, and female trees will then be preferred over the males!

Cutting Method

Grow your cutting in a nursery bed for a year or so to increase the tree's chances of survival when you transplant it to its final home.

1. Using diagonal cuts, take 4- to 5-inch tip cuttings from half-ripe wood (called a "softwood cutting") in July or August. With a sharp knife, cut the leaves in half to reduce transpiration, or water loss.

2. Dip the sliced end of each cutting in rooting powder, then place it in a container filled with a 50/50 mixture of sharp sand and peat moss.

3. Tent a piece of plastic over each cutting and its container. The plastic will create humidity, which encourages growth. Poke a few pencil-size holes in the plastic to allow a little ventilation.

4. Keep the medium moist but not wet until the cuttings begin to root. Then transplant the cuttings to a sandy loam bed that has been fortified with a good compost.

Sowing Seed

You should wear rubber gloves when you handle ginkgo seeds. Many people (about 50 percent, I am told) are allergic to one of ginkgo's constituents, called ginkgolic acid, which is similar to urushiol, the compound in poison ivy and poison oak that causes contact dermatitis.

1. Gather fallen ginkgo fruits in October and remove the outer fleshy layer, leaving the hard inner kernel. Wash the seeds well so that all of the fleshy material is removed.

2. Place the seeds in a box filled with moist sand, and store it in the refrigerator for 2 months or more.

3. Put the seeds into 4-inch pots containing a sandy loam that drains well. Keep the pots warm and moist until germination occurs. Plant out in the garden in light shade in the spring.

You may also want to direct sow in the garden in late fall. Young seedlings like a little shade, so pick a sheltered spot. As above, remove the outer fleshy layer from the seeds. Plant the seeds and cover them with 1 to 2 inches of sand or a sand/peat mixture. Cover with an inch of rotted leaves and wait for spring!

Alternative Seed-Germinating Method

I heard of a unique method for germinating seeds a few years ago, but I have not tried it myself. The source claims to get excellent germination. You might want to give it a shot.

After peeling and washing the nuts, put them in a barrel of water outdoors. I was told that any nuts that float should be removed because they will not germinate. Just let the nuts sit and freeze over the winter. In spring, drain off the water and plant the seeds as described above.

Grafting

Most plant nurseries use the grafting method to assure a crop of male trees. They take cuttings from male trees and graft them to young seedlings. It is also fairly common to graft a female branch onto a male tree so that pollination is assured but the volume of seeds is not overwhelming. Grafting is a complicated procedure that will probably not be of interest to smaller ginkgo growers, so I will not delve into those methods here. Rest assured, however, that there are many good books on the subject; if you would like more information, check your local library or bookstore for suggestions, or search the Internet.

Reproductive Anomalies

Yet another quirky thing about the ever-surprising ginkgo is that it is one of only two plants in the world that has motile sperm! (The other is the palmlike, tropical *Cycad* genus.) Amazingly, scientists are still not sure what goes on, and when it goes on, behind that fleshy outer skin of the ginkgo nut.

We do know one mind-boggling reason why ginkgo reproduces this way: It's so old that, at one time, there weren't any birds or even any insects on the earth to help with pollination! Can you get a feeling of just how incredibly old this tree species is?

The list of amazing things about the ginkgo goes on. Perhaps another reason why this tree has survived for so long is that, because of a backup method of propagation, it did not have to rely solely upon pollination to keep the species going and growing along the endless spectrum of time.

In addition to having motile sperm, some trees produce protrusions that are called *chi chi*, meaning *breasts*. These chi chi can grow all the way to the ground and form both roots and leaves, and thus another whole tree.

Growing Ginkgo as a Cash Crop

Ginkgo is one of the most popular medicinal herbs on the world market today. We may rightly infer that the hundreds of tons of ginkgo leaves in demand are not all harvested from trees growing in the wild. Ginkgos are now grown on large plantations containing millions of trees. Much of this supply comes from two plantations owned by the same company: Schwabe Pharmaceuticals.

The Birth of Garnay Ginkgo Plantation

Over 20 years ago, the German/French company Schwabe began growing ginkgo trees in the famous wine region of Bordeaux. The soil is rich and deep there, and the plantation found a rich market in Europe, where herbal medicine has a long, well-respected history. Ginkgo soon became one of the three top-selling products in France and Germany. Wanting to expand, the owners began looking for more land on which to grow the increasingly valuable trees.

Although the United States was still living in what I call the Dark Ages of herbal medicine, Schwabe successfully predicted that herbs would eventually become as accepted and utilized in North America as they are in European countries. Company personnel consulted with plant and soil scientists in the United States and found land in South Carolina that met their needs. Schwabe bought 1,150 acres of farmland there and began planting trees, creating the world's largest contiguous-acre ginkgo plantation. It was named Garnay, after the small French village where the founder of the company, Dr. Willmar Schwabe, had grown up.

Today, Garnay contains over 12 million trees. Each "tree" is only 4 feet tall and 2 feet wide. The trees are planted in very long rows so that they can be cultivated and harvested with specialized machinery.

Production at Garnay Plantation

Garnay trees are started from seeds obtained from Asia and sown in methyl bromide–treated nursery beds in April and May. After 2 years, when they are about 2½ feet tall, workers transplant the young trees to long rows in the growing fields. They space the trees far enough apart that tractors can navigate between the rows without damaging the plants. The young trees are fed, cultivated, and sheared.

Weed control is accomplished with a preemergent herbicide in early spring and an occasional direct-contact herbicide on emerged weeds. Workers also burn out weeds using propane torches attached to a brace at the front and side of a tractor that slowly moves down each row, burning weeds in the pathway as well as at the base of the ginkgo bushes.

In the summer of the third year, workers drive a modified cotton-picking machine down the rows. The machine twirls small, rubber batons that basically whip the leaves off the branches. At the same time, a huge vacuum sucks up the leaves as they are being whipped off the bushes and deposits them in a wire mesh–sided trailer that is towed by the tractor. When the trailer is full, workers bring it into a large steel building where the leaves are again vacuumed up, then deposited on a conveyor that moves them into a huge, barrel-like, gas-fired drying oven.

In only 8 minutes the leaves emerge fully dried and are dropped into a compactor, which presses the leaves into 300-pound bales that are bound in woven plastic cloth. The bales are forklifted into waiting trucks, driven to the docks at Charleston, and shipped to Ireland, where the processing takes place. At the Irish extraction facility, the leaves are mixed with acetone and water. After the right amount of time, the acetone is evaporated off and the dried extract is tested for residual chemicals and other undesirable substances.

For More Information

In addition to being a pioneer in large-scale growing of ginkgo, the Schwabe company also has contributed greatly to the scientific study of ginkgo. They have conducted numerous studies and clinical trials on the medicinal effects of ginkgo, which has both secured their leading market position and increased demand for their particular brand of standardized ginkgo extract, called EGb761. It is from Schwabe's studies that the preponderance of our present medical references are derived.

Ginkgo Production in China

During my recent visit to China, I was able to meet with the very knowledgeable professor of horticulture, Dr. He Shan-an, the former director of the Nanjing Botanical Gardens. He taught me much about the growing of ginkgo trees in China.

Although there are a few very large plantations, the majority of ginkgo leaves for the commercial trade are grown by small farmers, or farm cooperatives. Some of the trees are planted among vegetable crops for diversification, while other trees are grown in nurseries for landscape material.

Because there are no natural insects or diseases that are a major threat to this highly resistant (and resilient) tree crop, farmers use little, if any, pesticides or herbicides on the trees. However, Dr. He says the small farmers may spray the vegetable crops that grow around the ginkgo trees, thus subjecting the trees to either pesticide drift or systemic absorption through the roots.

Commercial Standardized Extracts

A standardized extract contains a measured amount of one or more of a plant's constituents. Standardized extracts are sometimes favored over whole-herb preparations (like teas or tinctures) because they allow measurable doses of what is thought to be the active constituent of an herb. However, as many herbalists point out, it may not be the "active" constituent that gives a plant its healing power but rather the balance of consituents, active or inactive, working together. The jury is still out on this question. Some people view standardization as a beneficial technological refinement of herbal medicine, while others view it as a profit-motivated aberration that takes folk remedies away from people and puts them into the laboratory and the hands of corporations and governments.

The Schwabe company, through its prolific research, set the standard for standardized extracts of ginkgo biloba. The usual strength of a standardized ginkgo capsule or tablet — standardized to contain 24 percent flavonol glycosides, 6 percent terpene lactones, and 3 percent ginkgolides — is 40 milligrams per pill. The usual daily consumption is three pills, though some people double this amount when dealing with serious health challenges. This brings up the very old question of whether more is better. If the medicine in an extract is more concentrated, is it stronger medicine? In some cases, for some people, this may be so. On the other hand, homeopathy is very effective for some people, and its healing premise is that the more dilute the remedy, the stronger it is. You must pay close attention to your body when taking remedies so that you can determine your own correct dosage.

Making & Using Your Own Ginkgo Tincture

The basics of herbal extraction are very simple. Plants contain within their cells both water-soluble and alcohol-soluble medicinal compounds. Teas, then, are water-based extracts of the water-soluble compounds; tinctures are alcohol-based extracts of the alcohol-soluble compounds. Since ginkgo is not particularly water soluble, it is more effective to extract your ginkgo leaves using the tincturing method.

> ## Handling Ginkgo
>
> As mentioned previously, some people develop a skin rash after handling ripe or rotting ginkgo fruit because of its ginkolic acid. However, I am unaware of any adverse reactions to harvesting the leaves at any stage of growth or to using a crude extract of ginkgo internally as a tea or tincture, other than gastric upset in a very small percentage of users.

Finding the Right Alcohol

Ethyl alcohol, also known as grain alcohol, in a concentration of at least 75 percent works best for tincturing. You can buy grain alcohol that is 150-proof (75 percent alcohol) in the form of rum. You may also be able to find pure grain alcohol (190-proof, or 95 percent pure alcohol), which you can dilute by 25 percent with water. Some states (including my own, Massachusetts) do not allow the sale of pure grain alcohol. If you can't buy grain alcohol or would prefer not to use such a high concentration, use 100-proof vodka (50 percent alcohol), which is easy to find.

Supplies and Equipment

To make a tincture of ginkgo, you will need the following supplies on hand:
- 200 grams (6 ounces) fresh ginkgo leaves
- 750 ml 100-proof vodka, or 600 ml pure grain alcohol diluted with 150 ml distilled water
- Electric kitchen blender
- Wide-mouthed jar that will hold at least half a gallon of liquid
- Bowl at least 8 inches wide
- Strainer (6 to 8 inches wide) or a potato ricer, cheesecloth, or cotton cloth
- Amber bottle with an airtight, polyseal top for storage (you can also use a sanitized beer or wine bottle)
- Amber or cobalt blue dropper bottles

All of the equipment you might need to tincture fresh ginkgo leaves can be found in your kitchen.

Step-by-Step Tincturing

1. Fill the blender with ginkgo leaves to within 1 inch of the top. Add either the vodka or the grain alcohol and water mixture to just below the level of the leaves.

2. Put the top on the blender and mix at medium speed for 10 to 15 seconds. Strain the liquid (which should now be bright green) into a bowl. Take the mash that is left in the strainer and put it into a wide-mouthed jar.

3. Refill the blender with more leaves. Pour the green liquid that is in the bowl back into the blender with the new leaves. Add enough vodka or alcohol/water mix to bring the liquid level to just below the level of the leaves again.

4. Blend as you did with the first batch, then strain the mixture and add the mash to the jar. Repeat steps 3 and 4 until you have used up all of the leaves, and all of the mashed leaves and alcohol are together in the wide-mouthed jar.

5. Seal the jar and set it in a cool, dark location to steep for at least 3 weeks. Shake the jar at least once a day, as agitation will enhance the extraction.

An out-of-the-way cabinet makes a good storage location for steeping tincture.

6. Strain and press the mash into a bowl using cheesecloth or a potato ricer. Funnel the tincture into a dark, glass, airtight bottle. This is your "master bottle" from which you can dispense tincture into 1-, 2-, or 4-ounce amber or cobalt blue bottles with eyedropper tops for easy measuring and administration.

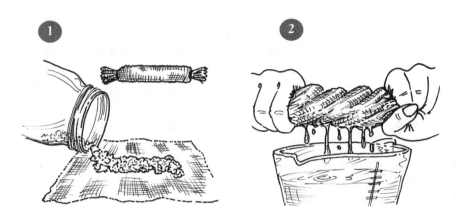

To drain every last drop of the tincture, pour the mashed herbs from the tincture onto a piece of cheesecloth. Roll up the cloth and wring out the precious liquid.

Be Sure to Label Your Bottle

Label your tincture with the following information:
- Name and weight of the herb (in this case ginkgo, 200 grams)
- Volume of alcohol (here it is 750 ml)
- Percentage of alcohol (100-proof liquor is 50 percent alcohol; the alcohol/water mixture described above is 75 percent alcohol)
- Date and "batch number" (if you make more than one)

You may also want to note where the herb came from: Did you buy it? If so, from whom? Or did you wildcraft it yourself? Where and when? This additional information allows you to compare the quality of various tinctures you make.

Ginkgo Tincture Dosages

The traditional amount of ginkgo tincture to take as an herbal supplement might be one dropperful, or about 30 to 35 drops, three to five times per day. You will find the taste somewhat bitter. Because a bitter taste helps to stimulate the secretion of bile and digestive juices and thus improve the digestive process, it might be wise to take your ginkgo just before meals. If you happen to dislike the taste of ginkgo (or any other tincture, for that matter), simply add it to your favorite beverage to mask the taste. When you take it regularly, you'll find that ginkgo is wonderful for enlivening and stabilizing the mind, eyes, ears, and nervous system.

Possible Side Effects and Contraindications

Ginkgo is an extremely friendly herb. It is no more dangerous than tomatoes or peanuts, to which some people can have an unpleasant reaction. In studies involving thousands of people, the most common negative side effects of ginkgo were gastric upset, nausea, diarrhea, edema, erythema, irritability, restlessness, and vomiting, which occurred in a miniscule percentage of the participants. These side effects cleared up upon discontinuation of the dosage.

In rare cases, people have experienced headaches after taking ginkgo. This may have happened because ginkgo did exactly what it was supposed to — increased cerebral circulation. The increased flow may have built up pressure in vessels unaccustomed to such a bounty of blood, food, and oxygen. If you experience a headache after taking the recommended dosage of ginkgo, reduce the dosage immediately, then increase it slowly over several days or weeks. If symptoms persist, discontinue using ginkgo, as you may be having a unique response to one or more of ginkgo's constituent substances.

If you are using a blood thinner or other pharmaceutical heart medication, consult a health care professional before taking ginkgo. There is no clinical information available on using ginkgo during pregnancy or while breast feeding, so consult your doctor or midwife before beginning to take it as a regular supplement.

My Favorite Ginkgo Recipes

There are many ways to take your ginkgo medicine. For example, you can drop some tincture into a glass of juice, sip it in a cup of warm tea, or enjoy a hearty chicken stew spiked with ginkgo nuts. Here are a few of my favorite recipes

The 4 Gs Revitalizing Tonic

My favorite herbs to combine in a tincture formula are unquestionably the four Gs: ginkgo, gotu kola *(Centella asiatica)*, ginger *(Zingiber officinale),* and American ginseng *(Panax quinquefolium).* Gotu kola is a gorgeous, tropical ground cover–type plant that, like ginkgo, can facilitate cerebral circulation. In northern climates it can be grown in the garden as an annual. Very much like the strawberry plant, it sends a tendril along the ground that then roots and starts another plant. From only one potted plant — divided and set out into the garden after any threat of frost has passed — it can cover a 10-foot-square area with a thick mat by early autumn.

Ginseng is a wonderful tonic revered throughout the world for its energizing properties. And ginger has long been used as a circulation stimulant, thus assisting (or "conducting," as it is described in traditional Chinese medicine) the other herbs in the formula.

To the 4 Gs core formula I add wood betony *(Stachys betonica).* I have long felt that this pretty, sun-loving perennial is one of my personal "plant allies" — a concept that anyone who has a deep love of medicinal herbs will understand. A well-known herbalist friend, Matthew Wood, says that this plant was used by early naturopaths for many of the conditions that ginkgo is used for today. I was delighted to learn of this and so included it in my recipe.

Finally, to add a delicious flavor to the formula, I put in a little of my own fresh vanilla-bean extract and some pure vegetable glycerine for sweetness.

THE 4 GS

This recipe not only helps you to remember where you left your car keys and what you had for breakfast, it also puts a spring in your step, lead in your pencil, and is really yummy to boot! It's like sleep — you need to use it often for best results.

- 6 parts fresh ginkgo tincture
- 4 parts gotu kola tincture
- 2 parts American ginseng tincture (from fresh root, if possible)
- 1 part wood betony leaf tincture
- ½ part pure vegetable glycerine
- ¼ part ginger tincture (from dried root, if possible)
- ⅛ part vanilla extract (fresh, if possible, although store-bought is fine)

To make:

Put all the ingredients in a "master bottle" for storage, then decant the mixture into 1-, 2-, or 4-ounce amber or cobalt blue bottles with eyedropper tops for easy measuring and dosing.

To use:

Like sleep, you need to use this often for best results. Take 30 drops to ¼ teaspoon three to six times a day. Start with the smallest dose and increase if necessary.

Gingko Blend Tea

The art of herbal formulation aims to increase the effectiveness of individual herbs by combining them with others to create a synergistic energy, in which the whole is greater than the sum of its individual parts. Although ginkgo's constituents extract more easily in alcohol, they combine well in this synergistic tea, which I call Monomoy Morning. It is useful for allergies, asthma, and low energy. It incorporates Chinese ephedra *(Ephedra sinensis)*, or *ma huang*, a stimulant that has been used for 5,000 years of recorded Chinese medical history. Because ma huang is a stimulant herb, it should be used with caution by those who have high blood pressure or anxiety. Fennel seed, cinnamon bark, carob pods, and cardamom seed add a delicious flavor as well as a "warming" quality. Licorice root adds sweetness and nourishes the adrenal glands, which may become exhausted from helping the body to deal with asthma or chronic stress.

I must tell you that this herbal beverage tastes even more delicious if it is made by the half-gallon and allowed to sit in the fridge for a few days before drinking. The herbs seem to mellow and blend even more synergistically after infusing in the cold water. And as the tea continually infuses, it is bringing you more medicinal quality. The herbs will usually saturate and settle to the bottom of the container, so you may not have to strain your tea until the last cup or two.

MONOMOY MORNING

Use dried, "cut and sifted" grade herbs except where noted otherwise.

- 8 parts ginkgo leaf
- 8 parts Chinese ephedra (ma huang)
- 6 parts licorice root
- 5 parts fennel seed
- 4 parts cinnamon granules
- 2½ parts carob powder
- 2 parts cardamom seeds, crushed
- ¾ part clove powder

To make:

1. Combine all the dry ingredients in a jar or bowl. Put ½ gallon of springwater in a pot (preferably not aluminum or copper) and add 6 to 8 tablespoons (depending upon desired strength) of the herb mixture.

2. Bring the mixture to the edge of a boil, then turn down to "warm" heat if you have an electric stove, or to a barely visible blue flame if you have a gas stove.

3. Allow to simmer for 5 minutes, then turn off the heat. Strain off a cupful now, if you like. Let the rest sit and steep for several more hours until it is cool, then pour it, herbs and all, into a wide-mouthed jar. Store the tea in the refrigerator, where it will keep for 4 to 6 days.

To use:

The traditional dosage for herbal teas is a 6- to 8-ounce cup of tea three times per day. However, you must listen to your body and adjust your dosage according to what feels right for you. Don't forget that herbalism is as much an art as it is a science! You may wish to begin your protocol slowly and increase gradually to be sure that you tolerate your remedy well.

> **Recycle Your Herbs**
>
> When you've strained all the liquid out of Monomoy Morning tea, put the spent herbs in your compost for future garden use.

Cooking with Ginkgo Nuts

It is interesting to note that in China and Japan, it is not the ginkgo leaf but the dehusked and shelled seeds that are traditionally used for medicine and in cooking. Early Asian cooks discovered that by boiling the nuts in two changes of water, they could safely use the cooked meat inside the shell as an amendment to their amazingly diverse cuisine.

GINKGO CHICKEN STEW

I like to make this chicken stew in an electric cooking pot, like a Crock-Pot, or on a wood stove. Some folks prefer to use a whole chicken, but I like to buy a package of leg quarters, which are both comparatively inexpensive and, in my opinion, more moist than the breast meat.

- 2 to 3 quarts water
- 5 carrots, chopped
- 5 celery stalks, chopped
- 2 heads broccoli (including stalks), chopped
- 1 head cauliflower (including stalks), chopped
- ½ medium-size winter squash, chopped
- 2 sweet potatoes, chopped
- 6 chicken leg quarters
- Salt and freshly ground black pepper
- 20 ginkgo nuts
- 1 tablespoon olive oil
- 1 large onion, diced
- 6 garlic cloves, minced
- 1 1" round fresh ginger root, finely chopped

To make:

1. In a large pot or crock, combine the water, carrots, celery, broccoli, cauliflower, squash, sweet potatoes, chicken, and salt and pepper. Bring to a simmer.

2. Put the ginkgo nuts (the cleaned, hard, pistachio-like inner seed without the outer fleshy layer) into a muslin teabag or a piece of tied cheesecloth and add it to the pot. Simmer the stew for at least 4 hours, or all day.

3. Just before the stew is cooked, remove the bag of ginkgo nuts and cool it under running water. Open the nuts and remove their meat. At the end of the kernel is a small piece like that on the end of a peanut. Remove and discard it.

4. Heat the olive oil in a pan and sauté the ginkgo nut meats, onions, garlic, and ginger. When this mixture is cooked to your preference, add it to the stew, call me for dinner, and serve. Yum!

Immunity-Enhancing Variation

To make an immune-enhancing variation of the Ginkgo Chicken recipe, add 6 sticks each of Chinese astragalus (*Astragalus membranacus*) and codonopsis (*Codonopsis pilosula*), also known as dang shen, to the pot when cooking begins. Both of these herbs are excellent for nourishing and building the deep immune system, and they are sweet and delicious. The codonopsis will become soft enough to eat along with the other vegetables. The astragalus can be chewed at the end of the meal, but it will be too woody to swallow. This stew is especially beneficial in the late fall and throughout the winter as a defense against colds, flus, bronchitis, pneumonia, and other seasonal maladies.

GINKGO NUTS TO MAKE BEAUTIFUL LADY

While I was riding on the train from Shanghai to Nanjing, a very nice and beautiful young college student asked the man next to me if she could have his seat so that she could practice her English with me. The man kindly obliged, and thus began a wonderful conversation. "Jane" (her chosen English name) told me that Chinese women use ginkgo nuts to make themselves beautiful. She described a recipe using broken English and pictures drawn in my travel log. I thought it so adorable that I have reproduced her very personalized recipe here.

To make:

1. Put your ginkgo nuts (the hard, pistachio-like seed) into a pot and cover with a few inches of water. Bring to a boil, then turn down the heat ("Fire not too big, not too small") so that the water is at the edge of a boil. Simmer until the water is almost all evaporated. Be careful not to burn the pot!

2. Let the nuts cool.

To use:

Eat six nuts each day — "not more than that!" says Jane.

Beauty Begins with Healthy Skin

"To be beautiful, first improve the diet," Jane instructed. As most herbalists will tell you, the quality of the skin is greatly dependent upon optimal functioning of the liver and kidneys, which help to eliminate natural metabolic toxins from the body. When these organs are in good shape, the burden upon the skin — our largest organ of elimination — to assist in detoxification is lessened.

In that ginkgo has, through five thousand years of recorded traditional Chinese medicine, been believed to have a beneficial effect upon the liver and kidneys, the connection between the use of ginkgo nuts and womens' beauty is quite rational. Treat the external by treating the internal. Beauty inside makes for beauty outside.

Ginkgo Nets to Make Beautiful Lady

① water — nets (with cover on) — Fire (not too big not too small)

② boiling.

③ water go out / leave only the nets / stop.

④ oooo net — take inside of nets / eat it as any other nut

To be beautiful first improve the diet.

6 (six) nets a day
not more than that.

曹琳曦

Conclusion

Ginkgo biloba is a truly amazing plant that, to me, represents all that ever was and gives us hope for all that ever will be. I thank you for reading this little collection of thoughts and enthusiasms and for sharing your own excitement and curiosity for the world's oldest tree. I hope to hear from some of you with stories, thoughts, and research of your own. You can subscribe to my quarterly newsletter called *News from Ginkgo Gulch*.

My final wish is that you remember that this planet and its intricate web of beings is perhaps not as resilient and strong as is the mighty and beautiful ginkgo. We must tread lightly, be conscious of our actions, and do what we are able in order to honor, help, and heal one another.

Love health and one another — Stephan

Resources

Great Cape Cod Organic Medicinal Herb Farm
P.O. Box 1206
Brewster, MA 02631
Phone: (508) 896-5900 or (800) 427 7144
E-mail: ginkgo@greatcape.com
Web site: www.greatcape.com
Specializes in medicinal herbs, herbal products, plants, seeds, and ginkgo trees.

Johnny's Seeds
1 Foss Hill Rd.
Albion, Maine 04910
Phone: (207) 437-4395
Supplies ginkgo seeds and small ginkgo seedlings.

Horizon Herb Seed Co.
P.O. Box 69
Williams, OR 97544
Phone: (541) 846-6704
Supplies ginkgo seeds.

CONVERTING MEASUREMENTS TO METRIC

Use the following chart for converting U.S. measurements to metric. Since these conversions are not exact, it's important to convert the measurements for all of the ingredients to maintain the same proportions as the original recipe.

To convert to	From	Multiply by
milliliters	teaspoons	4.93
milliliters	tablespoons	14.79
milliliters	fluid ounces	29.57
milliliters	cups	236.59
liters	cups	0.236
grams	ounces	28.35

To convert a Fahrenheit temperature to centigrade, subtract 32, multiply by 5, then divide by 9.

Other Storey Books You May Enjoy

Growing 101 Herbs That Heal: Gardening, Techniques, Recipes, and Remedies, by Tammi Hartung. Herb grower and herbalist Tammi Harting offers in-depth profiles for growing 101 medicinal plants using totally organic techniques. Hartung shares all the secrets of propagation, soil preparation, natural pest management, harvesting, and even garden design for both beauty and highest yield. Readers will learn to make inexpensive, potent home remedies and recipes for the whole family by following Hartung's easy, step-by-step instructions. Paperback. Full color. 256 pages. ISBN 1-58017-215-6.

Herbal Antibiotics, by Stephen Harrod Buhner. This book presents all the current information about antibiotic-resistant microbes and the herbs that are most effective in fighting them. Readers will also find detailed, step-by-step instructions for making and using herbal infusions, tinctures, teas, and salves to treat various types of infections. 144 pages. Paperback. ISBN 1-58017-148-6.

The Herbal Home Remedy Book: Simple Recipes for Tinctures, Teas, Salves, Tonics, and Syrups, by Joyce A. Wardwell. Enables the reader to identify and use 25 easy-to-find herbs to make simple remedies in the form of teas, tinctures, salves, tonics, vinegars, syrups, and lozenges. Gives hundreds of suggestions for maintaining health and well-being simply, naturally, and inexpensively. Folklore — including Native American legends and stories — provides information on the origins of many herbal medicines. 176 pages. Paperback. ISBN 1-58017-016-1.

Rosemary Gladstar's Herbs for Longevity and Well-Being, by Rosemary Gladstar. A thorough exploration of the life-extending properties of herbs such as gingko, ginseng, and echinacea, and their use in cultures around the world. Includes recipes to enhance quality of life, and tips for extending life through healthy living. 80 pages. Paperback. ISBN 1-58017-154-0.

Saw Palmetto for Men & Women, by David Winston. Respected herbalist David Winston brings a new perspective to using this popular herb for both men's and women's health problems such as prostate enlargement, male baldness, ovarian pain and cysts, infertility, cystic acne, anorexia, and as a booster for the immune system. 128 pages. Paperback. ISBN 1-58017-206-7.